Clock Repairing as a Hobby

CLOCK REPAIRING
AS A HOBBY

Harold C. Kelly

Association Press
New Century Publishers, Inc.

Clock Repairing as a Hobby

Printing Code

17   18   19

Library of Congress Catalog Card Number: 81-85509

ISBN 0-8329-1118-6

Printed in United States of America

Contents

Preface

This book is written for those who have a fascination for clocks but, even so, have never taken a look at the movement. Therefore, in analyzing the movement, the writer has kept in mind the questions a beginner is most likely to ask. Thus it follows that the first several chapters deal with movement design, escapements, pendulums, and balance wheels.

Although this book is basically a manual on clock repairing, it keeps its focus on the interests of the hobbyist, for there is no field in which one may find a more instructive and delightful hobby. The reasons for this approach are obvious when note is taken of the extensive literature available. There are many books covering the history of clockmaking. Others approach the subject as an art and many adherents become collectors. Still other books deal with clocks as a science in which physics and higher mathematics are expanded in great detail.

Thus it is seen that a clock hobbyist can expand his interest in a field that could become a lifelong study.

Part I

The Clock Movement—
Theory and Design

1

A Look at the Clock Movement

Clockmaking as a hobby should start with an understanding of the clock movement as a complete unit. This includes the source of power, the wheels and their purpose, and the mechanism that causes the pendulum to swing or the balance to turn.

This requires an understanding of the terminology used by the clockmaker. Consider first a number of definitions essential to this initial requirement.

Terminology

Wheel: any circular piece of metal on the periphery of which teeth may be cut of various forms and numbers.

Pinion: the smaller wheel with teeth called leaves that is driven by the wheel.

Barrel: (see Figure 46, p. 109) a circular box of metal for the reception of the mainspring.

Train: a combination of two or more wheels and pinions, geared together and transmitting power from one part of a mechanism to another part.

Main train: the train indicating the movement of time. The barrel is the first wheel of the main train.

Escape wheel: (see Figure 2, p. 14) the last wheel in the main train. The power residing at its circumference causes the escapement to function.

Escapement: (see Figures 5-12, pp. 21-34) those parts of the movement which change the circular force of the escape wheel into the vibrating motion of the pendulum or balance.

Pendulum: (see Figure 1, p. 13) a body suspended from a fixed point so as to swing to and fro; the time controller of a clock.

Balance: the wheel, which, in connection with a hairspring, functions as the time controller of a portable clock or watch.

Dial train: (see Figure 3, p. 15) a twelve-to-one gearing for the driving of the hour hand from the minute hand. Sometimes called motion work.

The Regulator Clock

The popular regulator clock is shown in Figure 1. Clock and watch repairers of the old school used the regulator for the testing and regulating of the clocks and watches brought in to them for repair.

The many parts that make up the movement may be separated into four separate mechanisms. These are: (1) the driving mechanism, (2) the transmitting mechanism, (3) the controlling mechanism, and (4) the time-indicating mechanism. The driving mechanism is the power plant and consists of a toothed wheel, a drum upon which a cable is wound, a click and ratchet to permit winding, and a weight. The transmitting mechanism carries this power through a

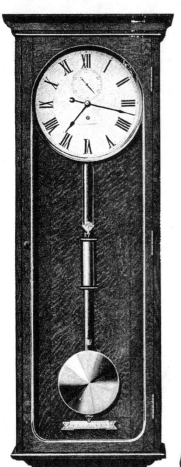

Figure 1.
American regulator clock

CLICK

RATCHET

DIAL TRAIN

DRUM

system of wheels and pinions and finally to the escape wheel. The controlling mechanism limits the turning of the escape wheel to a fixed number of turns per minute. This is accomplished through the action of the escapement and the pendulum as shown in Figure 2. The time-indicating mechanism consists of a dial, the hands, and the dial train. A

Figure 2.
Regulator movement showing the main train.
F—first or great wheel   T—third wheel
C—second or center wheel   E—escape wheel.
The small letters c, t, and e indicate
the pinions

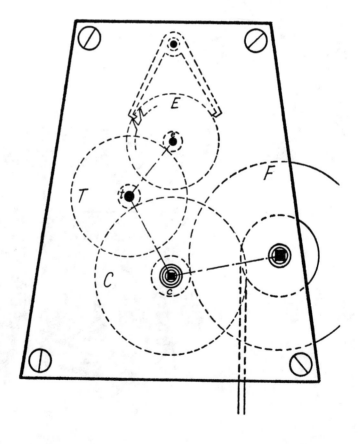

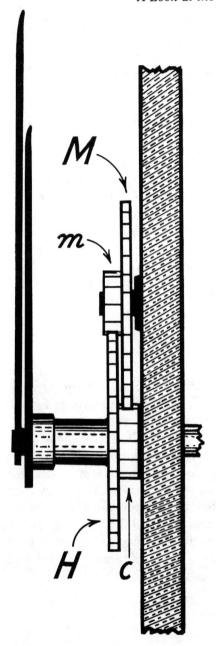

Figure 3.
Dial train.
c—cannon pinion
M—minute wheel
m—minute pinion
H—hour wheel

side view of the dial train is shown in Figure 3. This is the type of dial train used in regulators, French clocks, portable clocks, and watches. The cannon pinion is fitted to the post of the center wheel with a slight friction so that the hands may be set to time.

A Popular Model of an American Clock

The design of a clock that is characteristically an American creation is shown in Figure 4. Not only was this design popular in Colonial times but it is still popular even to this day. The only change that has been made is in the movement. The oldest models of these used wooden wheels and were powered by a weight. Later models used brass movements and were powered by a mainspring. Most American shelf clocks use the recoil escapement as shown here in Figure 4 and in Figure 5 on page 21. Note that the dial train is fitted between the plates. The necessary friction is provided by a three-prong spring that lies on the back side of the cannon pinion.

Portable Clocks

Portable clocks obviously cannot use the pendulum. In these the timing unit is a balance and hairspring. Also, the escapement must follow a different design. Figure 10 on page 31 shows the club-tooth lever escapement used in good grade portable clocks and watches. Figure 11 on page 33 shows the pin lever type used in alarm clocks and popular-priced wall clocks and watches. Also shown are two types of balance wheels used in portable timepieces.

Since the escapement, the pendulum, and the balance are the fundamental parts for the accurate registering of time, it follows that their action should receive special study and comment. This is given in Chapters 2, 3 and 4.

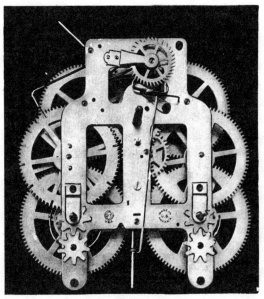

Figure 4.
American shelf clock

2

Escapements for Pendulum Clocks

The escapement for a pendulum clock is a mechanism for the purpose of maintaining the swinging of the pendulum. Conversely, the pendulum regulates the circular motion of the escape wheel to a required velocity. That type of escapement which interferes the least with the free motion of the pendulum will keep the most accurate time.

Terminology

A study of the escapement construction and action involves the use of many special terms that must be defined and understood before we can proceed with the analysis of clock escapements. The parts of the escapement for pendulum clocks are defined as follows:

Escape wheel: (see E, Figure 5, p. 21) the wheel of the main train that delivers impulse to the pendulum through the medium of the pallets and crutch.

Pallets: (see R and D, Figure 5, p. 21) that part of the escapement that receives the impulse from the escape wheel.

Crutch: (see D, Figure 37, p. 81) a slender metal piece that connects the pallets with the pendulum rod.

Receiving pallet: (see R, Figure 5, p. 21) the pallet over which a tooth of the escape wheel slides in order to enter between the pallets.

Discharging pallet: (see D, Figure 5, p. 21) the pallet over which a tooth of the escape wheel slides in order to leave from between the pallets.

Locking face: (see L, Figure 6, p. 23) the face of a pallet on which a tooth locks. This is a function that takes place only in dead-beat escapements.

Impulse face: (see I, Figure 6, p. 23) the lifting plane of a pallet.

Letting off corner: (see LO, Figure 6, p. 23) the extreme end of the impulse face of a pallet where a tooth of the escape wheel drops off.

Pendulum rod: (see G, Figure 37, p. 81) the connecting link between the suspension spring and the ball.

Suspension spring: (see G, Figure 37, p. 81) a short flat spring fitted to the top end of the pendulum rod. The suspension spring permits the pendulum assembly to swing to and fro.

Recoil Escapement

The recoil escapement, shown in Figure 5, was invented by Dr. Robert Hooke in 1671. It is sometimes referred to as the anchor escapement because its shape resembles that of an anchor.

Construction.   The escape wheel is of brass and the number of teeth varies greatly, depending on the type of clock and the length of the pendulum. The pallets are made of steel, hardened and highly polished.

Action of the escapement.   The pendulum is swinging to

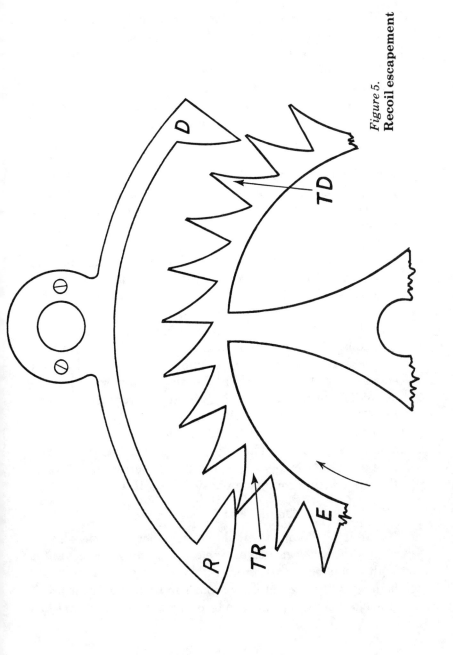

Figure 5.
Recoil escapement

the left, which produces a clockwise turning of the pallets. The tooth of the escape wheel **TR** delivers an impulse to the receiving pallet **R** and, following the impulse, the tooth drops off the pallet. The escape wheel is now free to turn, but its motion is immediately arrested because of contact of tooth **TD** with the discharging pallet **D**. At this instant the "tick" is heard. After the pendulum completes its outward swing, it stops momentarily and swings to the right. Again a tooth of the escape wheel delivers an impulse, this time by way of the discharging pallet. Eventually the discharging pallet releases a tooth and another tooth engages the receiving pallet, at which time the "tock" is heard. Thus the action continues as long as the proper motive power is maintained.

The recoil escapement gets its name from the fact that the escape wheel recoils or reverses its direction immediately after a tooth drops on a pallet. This action takes place because the pallets have no locking faces. Hence a tooth falls directly on the curved impulse face and, as the pendulum continues to swing outward, the curved pallet pushes the tooth back until the pendulum completes a swing and starts in the reverse direction. The actual impulse to the pendulum is found by subtracting the recoil from the forward motion of the escape wheel

The recoil escapement is still popular in the manufacture of low-priced clocks. The reason, no doubt, lies in the fact that it can be cheaply made and that it performs well even though not conforming to precision designing.

Dead-Beat Escapement

George Graham invented the dead-beat escapement in 1715. It is still regarded as one of the best escapements for high-grade regulators, grandfather clocks and others of the better quality. Except by comparison with the finest precision clocks, it leaves little to be desired. The only defect lies

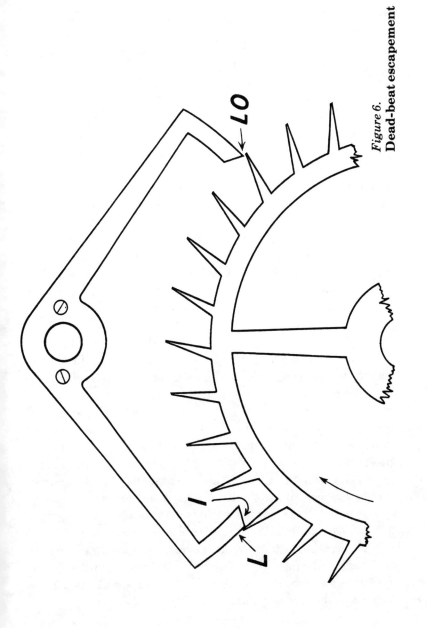

Figure 6.
Dead-beat escapement

in a slight tendency to change the rate of running after the oil has thickened or deteriorated.

Construction.   The Graham escapement is usually designed to embrace seven teeth of the escape wheel (Figure 6), although a design spanning ten teeth (Figure 7) has been used for large clocks and astronomical regulators. When the lesser number of teeth are embraced, the length and angle of the impulse faces are reduced and likewise the run on the locking faces. These factors become clear when comparing Figure 6 with Figure 7.

Action of the escapement.   The action of the dead-beat escapement is similar to the recoil escapement described above. One point of difference, however, should be noted. When examining the action, make certain that the teeth of the escape wheel drop on the locking faces and not on the impulse faces. This error could be due to worn pivot holes, for the dead-beat escapement calls for greater precision than is required in the recoil escapement.

Pin-Wheel Escapement

The pin-wheel escapement is occasionally found in regulator clocks and tower clocks. It was invented by J. A. Lepaute in 1750.

Construction.   Figure 8 shows the pin-wheel escapement. Semicircular pins stand out from the face of the escape wheel and engage the pallets on one side only, so that the pressure is always downward. The impulse is divided between the pins and the pallets. The amount of lock is determined by spreading or closing the pallet arms.

The pallet arms may be made to any reasonable length, a feature that is found in no other escapement.

The pin-wheel escapement has become obsolete because of manufacturing difficulties. It is more expensive to make and its performance is in no way superior to the Graham dead-beat escapement.

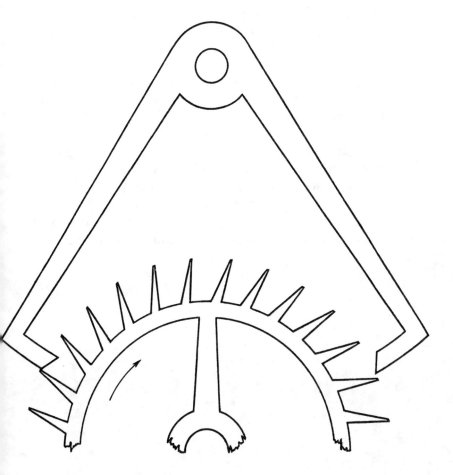

Figure 7.
Dead-beat escapement

Pin-Pallet Escapement

The pin-pallet escapement (Figure 9) may be found in the older American and French shelf clocks. The escapement often operates visibly in front of the dial, presenting an attractive appearance.

The pin-pallet escapement was invented by Achille Brocot about the year 1850 and is often referred to as the Brocot escapement.

Construction. The pin-pallet escapement has much in common with the Graham escapement—that is, its action is of the dead-beat type. Proportions vary, but the usual diameter of the pallets is a little less than the distance between two consecutive teeth of the escape wheel.

The pallets are usually jewels (carnelian, for example) and are held in place with shellac. If the jewels are lost or broken, pins made from steel rods will be satisfactory.

The pin-pallet escapement calls for hand assembly and adjustment in its manufacture. Thus, not being adapted to modern production methods, it has become obsolete.

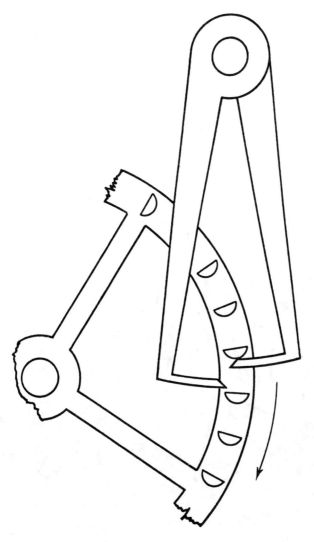

Figure 8.
Pin-wheel escapement

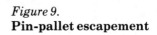

Figure 9.
Pin-pallet escapement

3

Escapements for
Balance Wheel Timepieces

The escapements described in Chapter 2 use
the pendulum as the time counter. Portable clocks and
watches require a balance and hairspring as the time counter and must of necessity use a different type of escapement.
This is called the lever escapement, and in the general run of
service we find three types. These are: (1) the club-tooth,
used in clocks and watches of the better quality, (2) the pin-
pallet, used in alarm clocks, wall clocks, and watches of low
cost, and (3) the ratchet-tooth. The ratchet-tooth type is
now obsolete, but is still seen in good quality clocks and
watches.

Club-Tooth Lever Escapement

Construction. The escape wheel of the club-tooth type
(Figure 10) is of brass or steel and usually has fifteen teeth
which act alternately on the receiving and discharging pal-

lets. The pallets of ruby or sapphire are cemented in slots in
the pallet frame. Impulse is divided between tooth and pallet. The pallet jewels are divided into three parts: locking
face, impulse face, and letting-off corner. The locking faces
of the pallets are tilted to the right in order to provide the
necessary draw. Draw may be defined as a force that holds
the lever against the banking pins.

Action of club-tooth escapement. Figure 10 shows a
tooth of the escape wheel impinging the receiving pallet.
The balance is rotating in the direction of the arrow, carrying with it the roller table. The roller jewel advances toward
the fork slot, contacts the right side of the slot, and moves
the lever to the right. This action causes the receiving pallet
to recede from the tooth until the unlocking takes place. Immediately after unlocking, the escape wheel turns clockwise and a tooth of the escape wheel, through contact with
the impulse face of the receiving pallet, drives the lever to
the right, and the fork slot carries the impulse to the roller
jewel. The force is sufficient to turn the balance with considerable velocity, unwinding the hairspring (in some cases
winding it up) until the force is exhausted. In the meantime
a tooth of the escape wheel has dropped off the receiving
pallet and another tooth has locked on the discharging pallet. The inertia of the balance having become exhausted,
the hairspring takes over and starts the balance on its return
vibration. When the roller jewel reaches the fork slot, the
discharging pallet unlocks a tooth of the escape wheel and
a new impulse begins, this time driving the balance in a
clockwise direction.

Beat.   The lever escapement is in beat when the hairspring is so adjusted as to permit the roller jewel to point toward the pallet center. In practice the balance may be turned an equal distance on either side of the line of centers,
released, and the watch should start. If it starts on one side
and sets on the other, the escapement is out of beat.

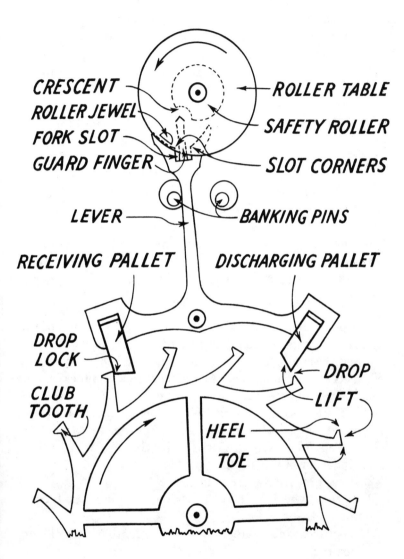

Figure 10.
Club-tooth lever escapement

Pin-Pallet Lever Escapement

Construction. The escape wheel of the pin-pallet type (Figure 11) is made of either brass or steel and usually contains fifteen teeth. The teeth are so shaped as to provide the entire lifting action on the pins **R** and **D**. The angle of the locking face also provides the function of draw. Except for the action of the escape wheel and pallets, the pin-pallet type functions in a manner comparable to the club-tooth type. So, additional comment is not needed.

Ratchet-Tooth Escapement

In the ratchet-tooth escapement (Figure 12), the entire impulse is provided by the pallets. Although this escapement is now obsolete, the ratchet-tooth type was popular in England for many years. This escapement is the oldest of the three designs mentioned above. It bears some resemblance to the lever escapement invented by Thomas Mudge in the year 1750.

The Tick and the Tock

The preceding analysis covers the general function of the lever escapement. One additional comment may well be of interest. This has to do with the ticking of the escapement. What is usually thought of as a two-beat tick and tock is really composed of five separate sounds which occur so close together that it is not possible to hear them individually. One sound with each balance vibration is all the ear can detect. The five sounds take place in the following order:

1. Roller jewel hitting the fork slot.
2. Fork slot hitting the roller jewel.
3. Escape wheel tooth starting down the impulse face of a pallet.
4. Tooth leaving a pallet and another tooth hitting the opposite pallet.
5. Lever hitting a banking pin.

N. B. The loudest ticks are numbers 1 and 4.

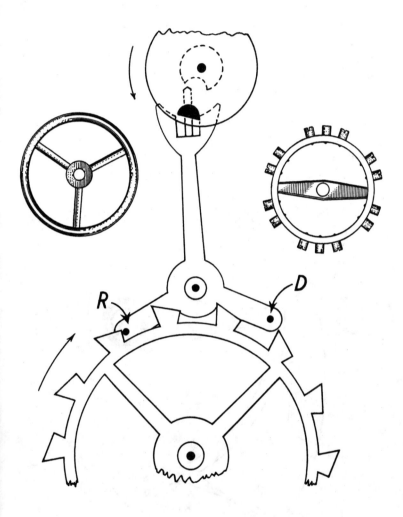

Figure 11.
Pin-pallet lever escapement

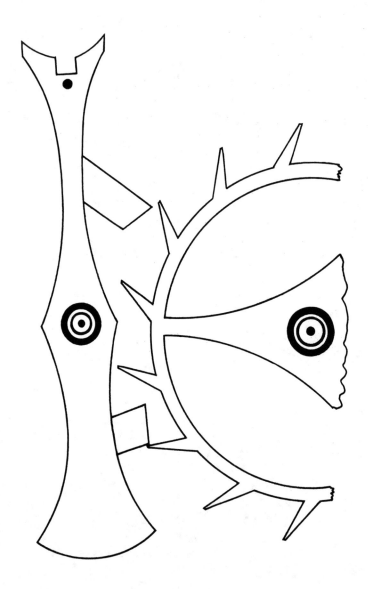

Figure 12.
English ratchet-tooth escapement

4

Pendulums

The invention of the pendulum, about the year 1657, ranks as one of the great inventions of all time. It revolutionized the construction of clocks. It is practically a precision instrument for the measurement of time. By means of a weight or mainspring, a train of wheels, and an escapement, its vibratory motion, which friction and the resistance of the air tend to destroy, is sustained.

As is the case with so many inventions, it is impossible to name any one person as the undisputed inventor. The two names most often mentioned as the original innovator, however, are Galileo Galilei and Christian Huyghens.

Motion of the Pendulum

A pendulum is in equilibrium when the rod which supports the ball is vertical. If the pendulum is pulled to an inclined position and let go, however, it will descend in order to regain its original position and thus its own momentum will cause it to rise nearly as far on the opposite side.

In its downward movement, the ball will move as far as the vertical position with an accelerated movement but not consistently so, as with a falling body, for the gravitational force along the circular path is continually diminishing and it ceases altogether when the ball reaches the vertical. On regaining its vertical position the pendulum will rise on the opposite side because of its acquired momentum. However, the effect of gravity which attracted the ball in its downward movement is now a resistance to its upward swing. The ball will, nevertheless, rise to nearly the same height on the opposite side and then return, executing swing after swing like the first.

Physical Phenomena Present in the Pendulum

Briefly stated there are three basic physical properties present in the pendulum. These are: (1) The fall of a body increases through the distance covered. Pendulums act accordingly and thus, according to the law, a pendulum one-fourth as long as another will swing twice as fast. (2) The velocity of a falling body is the same regardless of its weight. This means that altering the weight of a pendulum ball does not change the time of the swing as long as the center of oscillation is not changed. However, if a small weight is placed above the pendulum ball, the clock will run faster, for the center of oscillation has been raised and the pendulum has, in effect, been shortened. (3) When a pendulum clock is taken from one location to another in which the force of gravity is different, its rate is affected accordingly. Altitude is one factor in the force of gravity and centrifugal force (strongest at the equator) is another.

Detrimental Forces

There are several additional forces that alter the uniform

swing of the pendulum. Some can be quite accurately controlled, but others cannot. The more important sources of error are as follows:

1. Escapement error.
2. Circular error.
3. Temperature error.

These will now be discussed in the order named above.

Escapement error

The impulse to the pendulum through the escapement should take place at the moment when the pendulum is vertical. This ideal condition would permit the pendulum to swing to and fro in the same time that a free pendulum performs these swings. However, the mechanical means at our disposal to keep a pendulum swinging do not meet the above requirements and we are obliged to take account of the following laws:

1. An impulse delivered to the pendulum before the vertical point will accelerate the swinging.
2. An impulse delivered to the pendulum after the vertical point will retard the swinging.

These phenomena can be easily demonstrated with a pendulum. Impulse given to the pendulum before it reaches the vertical point causes it to arrive at the vertical more quickly than if it were acted upon by gravity alone. Impulse given after reaching the vertical results in driving the pendulum farther, resisting the force of gravity and at no particularly accelerated rate. Hence a retardation takes place, and the greater the distance the impulse takes place after the vertical, the greater is the retardation. This error also exists in balance wheel timepieces.

Circular error

Christian Huyghens, who built the first pendulum clock (Figure 13), discovered a problem that clockmakers to this

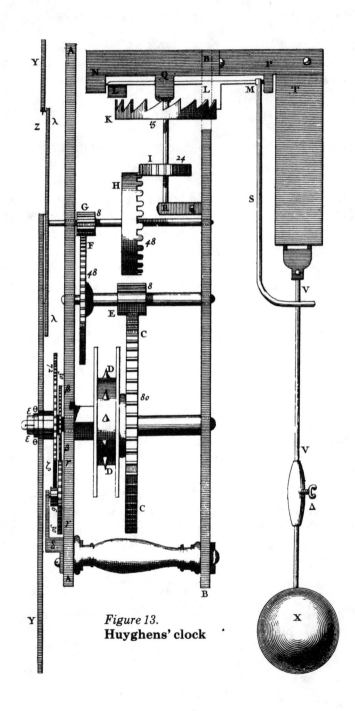

Figure 13.
Huyghens' clock

day have not solved. He observed that a pendulum making a long swing rates slower than when the same pendulum makes a short swing. Huyghens called this the circular error. The circular error was a real problem in Huyghens' clock, for the verge escapement of his day caused the pendulum to swing with exceedingly long arcs. Huyghens developed a solution by fitting metal pieces on each side of the suspension as shown in Figure 14. These metal pieces caused the pendulum to follow a steeper path near the extreme ends of the swing as shown in Figure 15. Huyghens' pendulum is not used with modern escapements because the pendulum arc is shorter and more consistently uniform, so that the circular error is materially reduced. The deterioration of the oil and the weakening of the mainspring as the clock runs down, however, tend to shorten the pendulum arc and the circular error then becomes manifest.

Temperature error

Most materials of which pendulums are made expand in heat and contract in cold. Therefore a clock without some sort of temperature compensation will run fast in winter and slow in summer.

The theory of temperature compensation as applied to pendulums is simply that the center of gravity of the ball is raised or lowered by an amount equal to the lengthening or shortening of the pendulum rod. This is done by means of the regulating nut at the bottom of the ball.

Wood for the pendulum rod and lead for the ball are suitable for domestic and grandfather clocks, but are not sufficiently accurate for precision clocks.

Graham's mercury pendulum.   George Graham invented the mercury pendulum in 1715 and it was the first successful attempt at temperature compensation with any degree of precision. His pendulum consisted of a steel rod and a

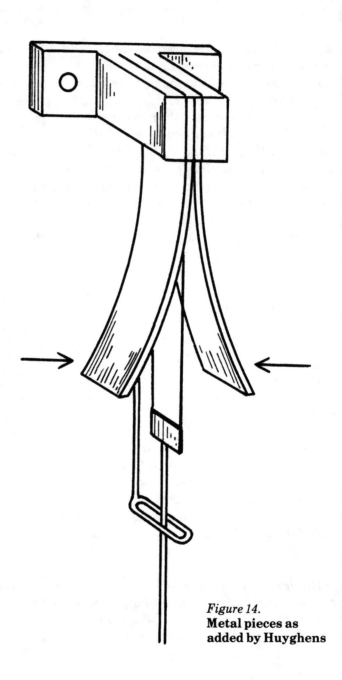

Figure 14.
**Metal pieces as
added by Huyghens**

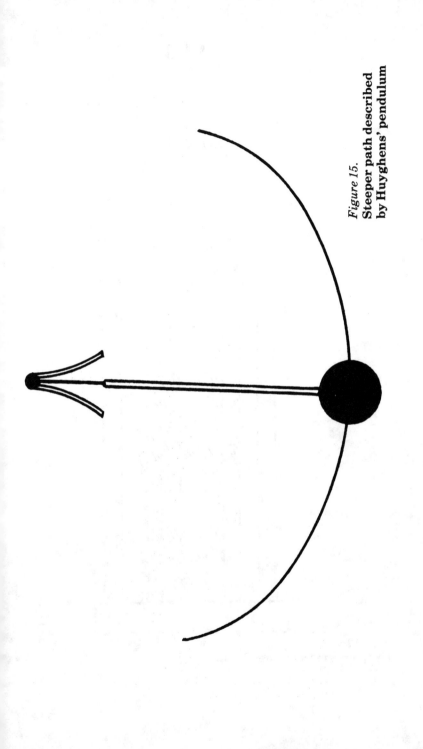

Figure 15.
**Steeper path described
by Huyghens' pendulum**

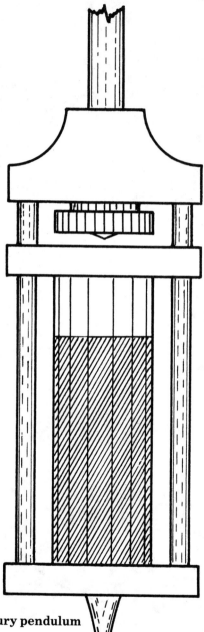

Figure 16.
Graham's mercury pendulum

single jar of mercury as shown in Figure 16. Two jars are sometimes used, with the advantage that the mercury responds to temperature fluctuations more quickly.

Mercury is an excellent compensating material since it is liquid and can be removed or added to the jar so as to attain perfect compensation. For example, if the clock loses in heat, the pendulum is said to be undercompensated and more mercury should be added.

Harrison's gridiron pendulum. This pendulum, invented by John Harrison in 1720, is shown in Figure 17. The ball is supported by five steel rods and four brass rods. The downward expansion of the steel rods is compensated by the upward expansion of the brass rods. A few clocks with the Harrison gridiron pendulums may still be seen in jewelry store windows, but this pendulum is now regarded as obsolete.

Invar pendulums. Pendulum rods made of Invar, an alloy of 64 parts of steel and 36 parts of nickel, have the curious property of little or no expansion. Invar, coined from the word "invariable," is the result of years of research by Dr. Charles E. Guillaume, late director of the International Bureau of Weights and Measures at Sèvres. The difficulties in the making of Invar are such that no two batches are exactly alike and each pendulum calls for individual experimental compensation.

The ball for an Invar pendulum is usually a steel cylinder (Figure 18), drilled lengthwise to admit the Invar rod. From the bottom to the center, the bore is larger so as to permit a space for a brass or steel pipe called the compensator, which provides the upward expansion. The regulating nut lies below the compensator and is screwed on a threaded portion of the Invar rod.

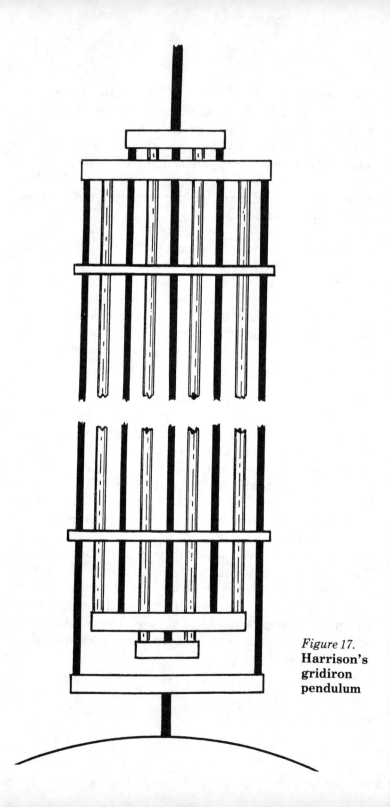

Figure 17.
Harrison's gridiron pendulum

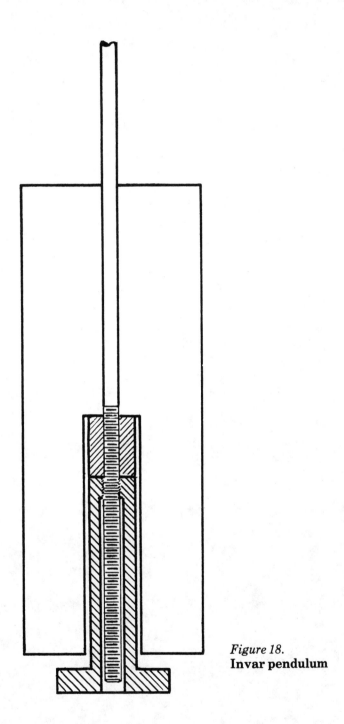

Figure 18.
Invar pendulum

5

Striking Mechanisms

The striking mechanism of a clock consists of a separate train and power unit. It is not in any way connected with the timekeeping mechanism, except that the main train sets the striking train in motion at the proper time.

There are two types of striking mechanisms in general use. These are the counting-wheel and the rack-and-snail. The counting-wheel is the older; in fact, it is as old as the mechanical clock itself. It is still used in American shelf clocks and in tower clocks. The rack-and-snail type was invented by Edward Barlow in 1676 and is found in shelf clocks of the better quality and in all modern grandfather clocks.

Counting-wheel Mechanism

The only disadvantage of the counting-wheel type is the caution required in setting the hands to time. In advancing the minute hand, it is necessary to allow the striking to finish to the hour indicated before advancing the hand to the next half hour. In other words, the counting-wheel system strikes consecutively and is not self-correcting.

American shelf clock with counting-wheel mechanism

Construction.   Figure 19 shows the arrangement usually found in the American shelf clock. The cam on the center arbor **CA** is shown in a position ready to raise the lifting lever **L**. This lever is attached to the arbor **A**, which, in turn, carries the warning lever **W** and the unlocking lever **U**. The unlocking lever lies under and pushes the counting lever **CL** which is secured to the arbor **B**. Mounted on arbor **B** are two levers: the counting lever, mentioned above, and the drop lever **D**. When the striking mechanism is locked, the drop lever **D** rests in the notch of the cam wheel **C**. Also, the counting lever **CL** rests in one of the deeper slots in the counting wheel **CW**.

Action of the counting-wheel mechanism.   As the cam on the center arbor **CA** advances clockwise and raises the lifting lever **L**, four levers are brought into action. Warning lever **W** is raised in position to engage the pin **P** on the warning wheel. The unlocking lever **U** lifts the drop lever **D** and also the counting lever **CL**, since both levers are attached to the arbor **B**. This results in a simultaneous separation of these two levers from their seats—the former from the notch in the cam wheel and the latter from the deeper slot in the counting wheel. The train is now free to turn but only for a moment, because pin **P** on the warning wheel will catch on the already raised warning lever **W**. Precisely on the hour, the cam on the center arbor **CA** passes the lifting lever **L**, permitting the lifting lever to fall. At the same moment, the warning lever **W** also drops. This action releases the pin **P** on the warning wheel and the striking begins. The speed of the striking mechanism is controlled by the Fly **F**. As the striking continues, the counting wheel turns, and the count-

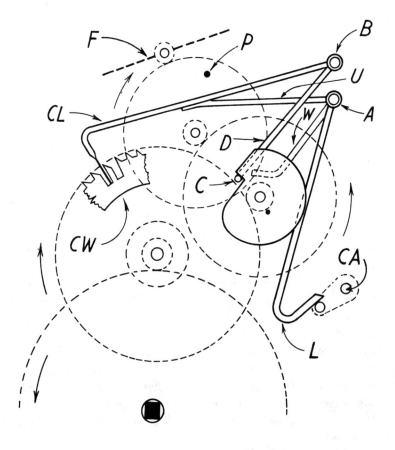

Figure 19.
**Counting-wheel striking mechanism
as applied to American shelf clocks**

ing lever drops into and raises above the shallow slots. Finally the counting lever drops into one of the deeper slots and the drop lever, at the same moment, drops into the notch on the cam wheel. When this happens, the striking stops.

Grandfather clock with counting-wheel mechanism

Construction. It will be noted that in Figure 20 the pin **WP** on the warning wheel **WW** performs two functions: (1) receiving the warning lever **WL** prior to the striking, and (2) receiving the drop lever **DL** to stop the striking. This differs from the counting-wheel system previously considered where the striking train was stopped when the drop lever dropped and locked in the slot in the cam wheel. It is apparent that the counting wheel and the cam wheel must be carefully timed in order to function correctly.

Action of the striking mechanism. The striking train is stopped by the pin **WP** on the warning wheel engaging the locking point **L**. This is possible because the notch in the cam is in position to receive the dropping point **D**, and the slot in the counting wheel is in position to receive the counting point **C**.

The dial train is so designed that the minute wheel **MW** turns once an hour in order that the lifting pin **LP**, which sets off the striking train, may be attached to it. The lifting pin advances in the direction of the arrow, and raises the lifting lever **LL** and likewise the warning lever **WL**. The pin **P**, attached to the warning lever, lifts the drop lever and releases the pin **WP** on the warning wheel. The warning wheel is now free to turn, making nearly one revolution. It is then stopped when the pin **WP** engages the warning lever **WL**.

On the hour, the lifting pin on the minute wheel passes the lifting lever **LL**, permitting it to drop. The warning lever drops accordingly and the striking begins. The drop lever does not now drop, since the cam wheel has turned a suffi-

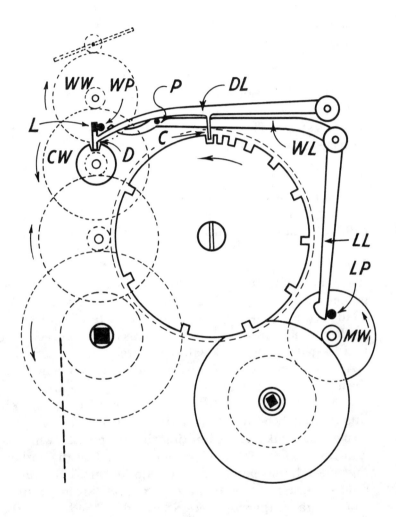

Figure 20.
**Counting-wheel striking mechanism
as applied to grandfather clocks**

cient distance to prevent it. Striking, therefore, continues until the counting point **C** drops into one of the slots on the counting wheel, and, at the same instant, the notch in the cam wheel **CW** comes into position to receive the dropping point **D**.

Rack-and-snail Mechanism

The rack-and-snail striking system differs from the counting-wheel type in that the hour wheel regulates the number of blows struck. This being the case, the striking is always corrected to the hour indicated.

Grandfather clock rack-and-snail mechanism

Construction.   Figure 21 shows the rack-and-snail striking system as applied to grandfather clocks. The disk **SN**, called the snail, is attached to the hour wheel and has twelve steps which govern the depth to which the rack **R** drops. The top portion of the rack usually has fifteen teeth and the number of teeth pulled to the right by the beak **B** of the gathering pallet **G** is determined by the position of the snail **SN**. Suppose, for example, the rack tail **TL** drops to the lowest step on the snail. Twelve teeth on the rack will be liberated and the clock will strike twelve.

Action of the rack-and-snail mechanism.   The pin **P** on the minute wheel advances in the direction of the arrow and raises the lifting piece **LP**. Then the warning lever **WL** lifts the rack hook **RH**, freeing the rack which falls to some step on the snail. At the same time the stop pin **SP** (which is attached to the rack) recedes from the tail **T** of the gathering pallet **G**, thus permitting the train to start. The train is soon stopped, however, because the pin **WP** on the warning wheel engages the stop piece **S** located on the back of the warning lever **WL**.

Later, on the hour, the pin **P** on the minute wheel passes the end of the lifting piece **LP**, permitting the warning lever

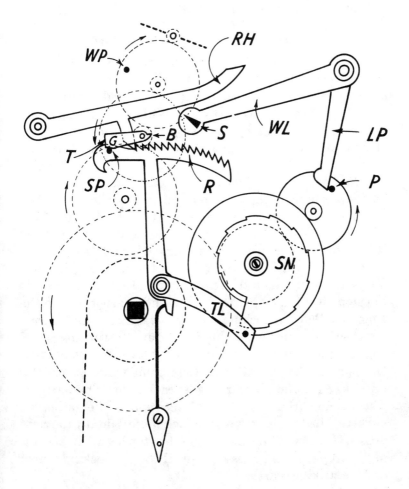

Figure 21.

**Rack-and-snail striking mechanism
as applied to grandfather clocks**

WL to drop and free the warning wheel. The striking train is now set in motion and the beak **B** on the gathering pallet **G** proceeds to collect the teeth on the rack. After the last tooth is collected, the stop pin **SP** advances in the path of the gathering pallet's tail **T** and the striking train stops.

Shelf clock rack-and-snail mechanism

Construction. The rack-and-snail system employed in shelf clocks differs slightly from that generally used in grandfather clocks. The locking of the striking train in weight-driven clocks occurs by means of a tail on the gathering pallet, which engages the stop pin on the left portion of the rack. This plan is not used in those movements driven by a mainspring, as the following description will show.

Action of shelf clock rack-and-snail mechanism. Figure 22 shows the French clock movement as typical of those found in most shelf clocks. A pin **MP-1** on the minute wheel raises the lifting lever **L**, carrying with it the warning lever **W**. The lifting lever also raises the rack hook **RH** by means of the pin **P** which in turn releases the rack **R**. The rack drops to some step on the snail **S**, and at the same instant the stop lever **SL** (fitted to the rack hook arbor) clears the stop pin **SP** on the locking wheel. The striking train turns, but is soon stopped when the pin **WP** on the warning wheel engages the stop piece on the warning lever **W**. Eventually, on the hour, the pin **MP-1** on the minute wheel passes the lifting lever **L** and the warning lever **W** drops. The warning wheel is now freed and the striking begins. The gathering pallet **G** collects the teeth on the rack **R** until finally the rack hook **RH** drops off the last tooth. The stop lever **SL**, accordingly, drops and engages the pin **SP** on the locking wheel and the striking ceases.

Half-hour striking. The usual method employed in half-hour striking is to make the first tooth on the rack **R** shorter than the rest and to fit a second pin **MP-2** on the minute

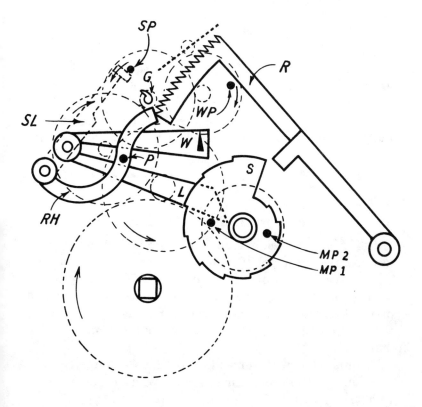

Figure 22.
**Rack-and-snail striking mechanism
as applied to shelf clocks**

wheel. The pin **MP-2** lies a little nearer to the center than the pin **MP-1**. In this way, the rack hook **RH** is lifted to clear the first tooth only, and the clock strikes one on the half hour.

Ship's Bell Strike

The ship's bell striking work usually employs the rack-and-snail system. The manner of striking differs, however, as the plan adopted is for the purpose of calling the crew to regular work shifts called "watches."

At 12:30 p.m., the clock strikes one, called one bell. At 1 p.m. it strikes two, or two bells; at 1:30 p.m., three or three bells, etc. Each half hour a bell is added until eight bells are struck at 4 p.m. The plan continues throughout the 24 hours, with eight bells being struck at 12 noon, 4 p.m., 8 p.m., 12 midnight, 4 a.m., and 8 a.m. After one bell the striking occurs in pairs as may be noted by the following example:

Chimes and Chiming Mechanisms

Sets of bells, struck by hammers and playing fascinating tunes, have been popular for centuries in clockmaking. There are a number of bell combinations, commonly called chimes. Most of these play at quarter-hour periods. Some play one chime, others play two or three. In the latter case, the desired chime is selected by setting a hand on the clock dial.

The more popular chimes include the Westminster, Wittington, and Guildford. These are shown in Figures 23, 24, and 25.

Chiming mechanism

Construction. Chiming clocks use three trains: (1) hour-striking train, (2) time train, and (3) chiming train. The time

Figure 23.
Westminster chime

Figure 24.
Wittington chime

Figure 25.
Guildford chime

train occupies the center of the movement, the hour-striking train is planted to the left, and the chiming train to the right. The rack-and-snail system is the one usually used. Occasionally one finds the counting-wheel used on the chiming train and the rack-and-snail on the hour-striking train. The mechanical principles vary somewhat in the different makes, but the majority follow the plan shown in Figure 26.

The chime snail **CS** is mounted on the minute wheel and carries four pins for raising the chime lifting lever **CL**. Also attached to the chime snail is a pin for raising the strike lifting lever **SL**. The strike warning lever **SW** releases the strike warning wheel **S** when the chime gathering pallet **G** has gathered up the last tooth on the chime rack.

Action of the chiming and striking mechanism. The pin **P**, on the chime snail, advancing in the direction of the

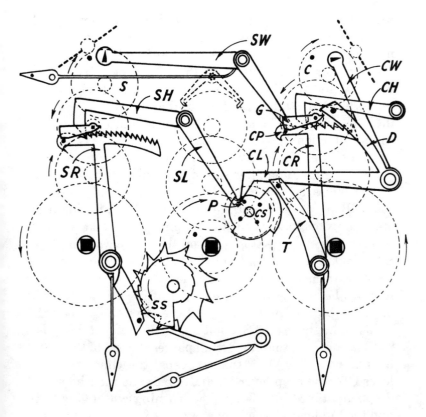

Figure 26.
**Chiming and striking mechanism.
Rack-and-snail system**

arrow, raises the chime lifting lever **CL**, and the chime warning lever **CW** raises with it. The chime rack hook **CH** is also lifted by means of the drop lever **D**, thereby releasing the chime rack **CR**. The tail **T** of the chime rack engages the largest radius of the chime snail **CS**, permitting the chime rack to drop a distance of one tooth. When the rack drops, the tail of the chime gathering pallet **G** is released from the rack pin **CP**. The train starts to turn, but is soon stopped when a pin on the chime warning wheel **C** engages the already raised chime warning lever **CW**. On the hour, pin **P** on the chime snail passes the end of the lifting lever **CL**, permitting the warning lever **CW** and the drop lever **D** to drop. The warning wheel **C** is freed; the rack hook **CH** drops to the rack and the first quarter chime is played. The chime ceases when the tail of the chime gathering pallet **G** engages the rack pin **CP**. The second and third quarters and hour chimes function in like manner.

Prior to the release of the chime rack at the hour chime, a pin on the chime snail has raised the strike lifting lever **SL**. The strike rack hook **SH** is likewise raised and the strike rack **SR** drops to the strike snail **SS**. The striking-train starts, only to be stopped when the pin on the strike warning wheel **S** engages the stop piece on the strike warning lever **SW**. Thus it is seen that both the chime warning and the strike warning take place before the hour chime plays. A few minutes before the hour, the strike lifting lever **SL** drops off a pin on the chime snail and the strike rack hook **SH** drops to the strike rack **SR**, but no further action takes place until the pin **P** on the chime snail passes the end of the chime lifting lever **CL**. When this happens, the chime warning lever **CW**, the drop lever **D**, and the chime rack hook **CH** drop simultaneously. The chime warning wheel **C** is freed and the full hour chime is played. As the last movement of the hour chime is playing, a pin on the chime rack pulls up the strike warning lever **SW**, releasing the strike warning wheel **S**. Thus the striking of the hour immediately follows the hour chime.

It will be observed that the strike warning lever is moved near the completion of each quarter chime, but no action takes place until such time that the strike rack has dropped.

Part II

Practical Clock Repairing

6

Tools, Materials,
and Supplies

When the horological student first examines the tool and material catalogs of those companies who cater to the needs of horologists, he is, no doubt, bewildered because of the vast array of tools, materials, and supplies. It is to be admitted that a complete set of tools runs into a considerable amount of money. The clock hobbyist need not be disturbed, however, because it is possible to start working with only a few essential tools. Later, as new techniques are learned, the necessary tools may be purchased. In fact, most horologists build up their tools and equipment gradually over a period of years. Horology, it seems, is that type of semiprofession in which the unfolding skills develop in order from one prerequisite to the next. This explains why the tool requirement is one of gradual expansion, but first let us consider the most essential tools.

Tools and Their Uses

Screwdrivers.   In clock repairing, three screwdrivers of varying sizes are needed. These are shown at **A**, **B**, and **C**, Figure 27. One screwdriver, shown at **C**, holds the screw and this is a real help in replacing the screws that hold the movement in the case.

Small screwdrivers for use on portable clocks may be purchased singly or in sets as shown at **A** and **B** in Figure 28.

After much use the blades of the screwdrivers become worn and need to be restored. This can be done on the hard Arkansas oilstone shown at **C** in Figure 28. A still better method is to use the lathe (Figure 29) and a carborundum wheel shown at **B** in Figure 29.

Tweezers.   The supply of tweezers should be quite large, as is seen in Figure 30. Long heavy tweezers are needed for the assembly of the larger clocks and smaller tweezers for portable clocks. One pair with fine points is reserved for hairsprings and should not be used for any other work. Tweezers, like screwdrivers, need to be reconditioned occasionally.

Eye loupes.   For general use a 3-inch eye loupe is needed (Figure 31). The hobbyist who has difficulty in holding an eye loupe may use the one with a wire attached. This wire is secured around one's head and this holds the eye loupe in place. If one wears eyeglasses, use the eye loupe shown at the right in Figure 31.

Other essential tools

Many uses will be found for two hammers; one with a steel head for riveting, and one of brass that will not mar the work. Four types of pliers are indispensable. The flat-nose type may be used for bending the crutch in order to put the escapement in beat. The striking, too, may be adjusted with the flat-nose pliers.

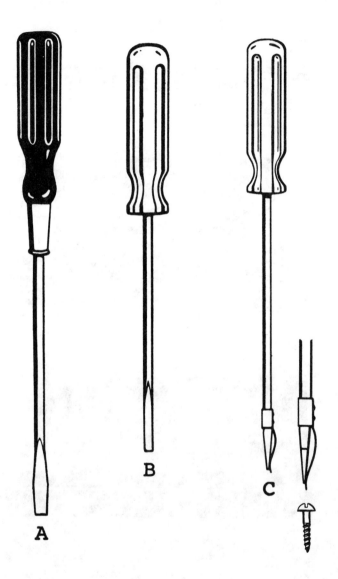

Figure 27.
Screwdrivers needed for clock repairing

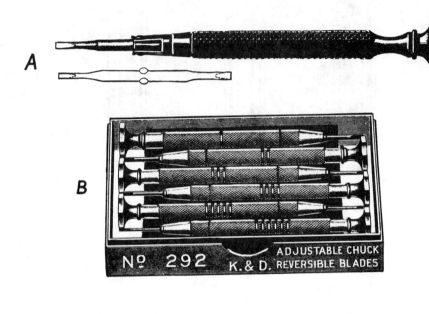

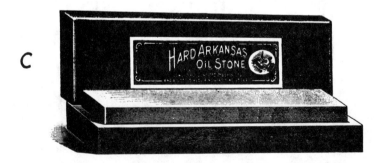

Figure 28.
**Small screwdrivers come singly or in sets.
Arkansas oilstone is handy for
renewing screwdriver blade**

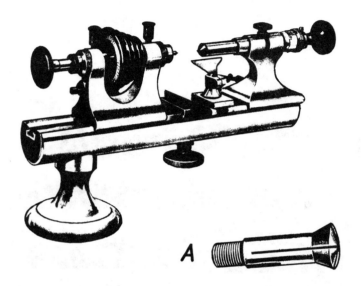

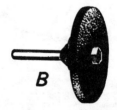

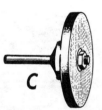

Figure 29.
Lathe and grinding wheels.
A—lathe chuck
B—carborundum wheel
C—Arkansas wheel

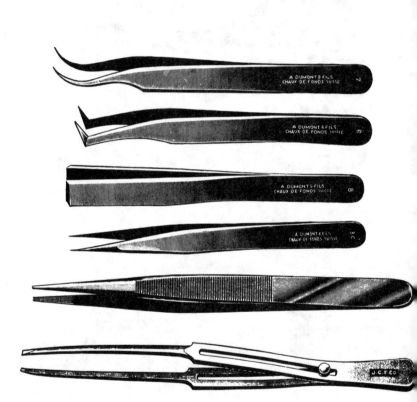

Figure 30.
Tweezers used in clock assembly

Figure 31.
Eye loupes

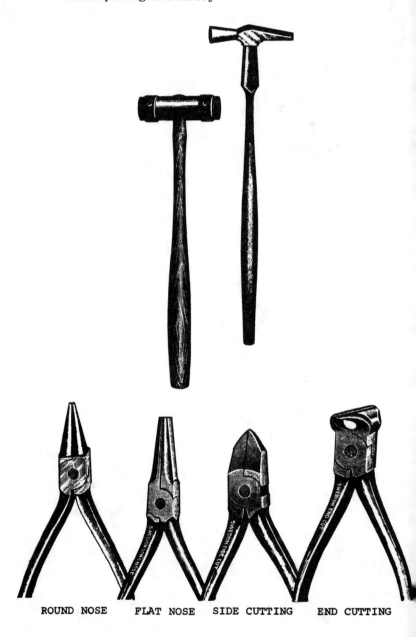

ROUND NOSE   FLAT NOSE   SIDE CUTTING   END CUTTING

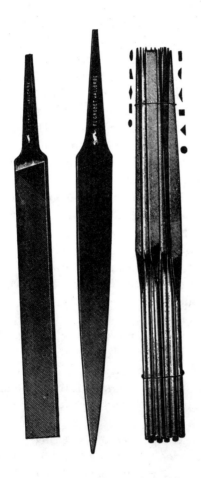

Figure 32.
Hammers, pliers, and files

Two files are needed for general use and a set of needle files for specialized use such as the shaping of wheel teeth. All these tools are shown in Figure 32.

Tools for special uses

There are many tools designed especially for the clock repairer and these are shown in Figures 33 through 36. **A, B,** and **C** in Figure 33 are hand removers for removing the hands in certain clocks. In most shelf clocks the minute hand is held in place by a nut and in others by a dial washer and a pin. After removing the nut or pin, the hands can usually be removed with the fingers. In some portable clocks, however, the minute hand is fitted friction tight and in these the hand remover is needed.

At **D** in Figure 33 is shown what is known as a bench key. They usually come in sets of two and are used to let the mainspring down prior to dismantling the movement.

The brushes shown at **E, F,** and **G** in Figure 33 are used in cleaning the movement. The brush shown at **F** is particularly useful in cleaning the area around the pivot holes. The one noted at **G** is designed to reach inside the movement.

Punches used for closing holes in clock plates are shown in Figure 34. The punches have spring centers that may be placed over the holes. The set of punches shown at **A** have spring centers with pointed tips, while the set of punches shown at **B** have spring centers with round ends. The small and large anvils shown at **C** and **D** provide working surfaces for the easy use of punches.

Figure 35 shows tools that are called pin vises and broaches. The pin vise serves as a handle to hold a broach. Broaches are used to enlarge holes. After worn holes in the plates are closed, broaches are used to make the holes larger so as to properly fit the pivots of the wheels. The larger broaches are used to make holes of a size needed to receive the bushings.

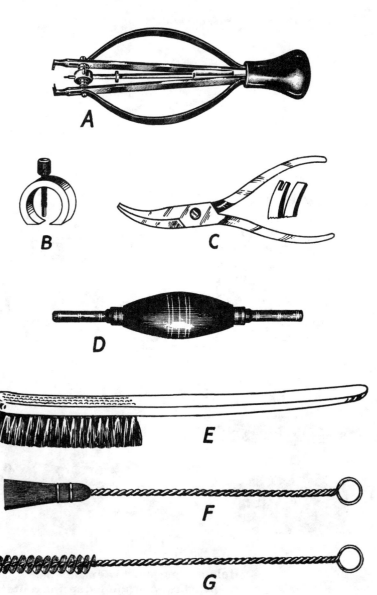

Figure 33.
Special tools.
A, B and C are hand removers
D is a bench key, used for unwinding mainsprings
E, F and G are clock cleaning brushes

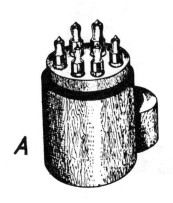
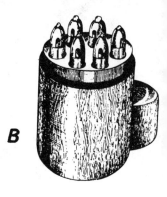

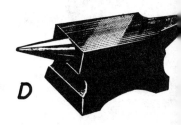

Figure 34.
Hole-closing punches and anvils.
A—punches with pointed spring centers
B—punches with round spring centers
C—small anvil
D—large heavy-duty anvil

Figure 36, reading across, shows (**A**) an Arkansas slip, (**B**) a Number Three emery buff, and (**C**) a pivot file and burnisher. The Arkansas slip and the Number Three emery buff are used to restore the polish on the impulse faces of the pallets. The pivot file and burnisher (the file is at the bottom, the burnisher is at the top) is used to repair and polish the wheel pivots. The technique for using the above tools is given in Chapter 7.

Materials and Supplies

It is desirable to have a supply of clock pins and clock washers, for these are sometimes missing in the movement or possibly may be lost in the course of repairing. These are supplied in assortments by tool and material dealers and come in small boxes as shown at **A** and **B** in Figure 37.

Clock bushings, too, are available in assortments. One bushing is shown at **C**.

Parts that occasionally need replacing are the escapement complete with the crutch (**D**); the pendulum rod with suspension spring (**G**); the pendulum ball (**E** and **F**); and quite frequently a mainspring (**H** and **I**). In many instances it is best to send the old part as a sample to the dealer when ordering a new part.

The materials and tools for cleaning and oiling the movement include cleaning and rinsing solutions, clock oil, oil cups, clock oilers, and pithwood. These are shown in Figure 38. If the fountain oiler, shown at **D**, is used, then oil cups are not needed. A quantity of oil is retained in the glass handle, and the oil is fed into the hollow applicator end by capillary action. When the tip of the needle is touched to the part needing oil, just the right amount of oil is deposited. The pithwood shown at **E** is used to keep the ends of the oilers clean.

Watch and clock repairers of the old school used gasoline, benzene, soap and water, or chalk to clean the movement.

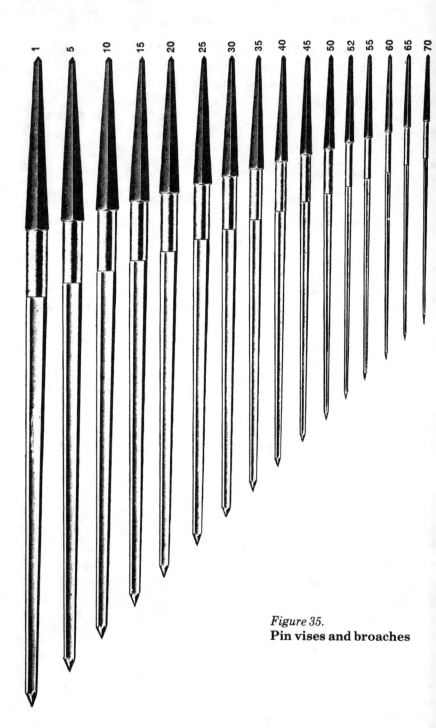

Figure 35.
Pin vises and broaches

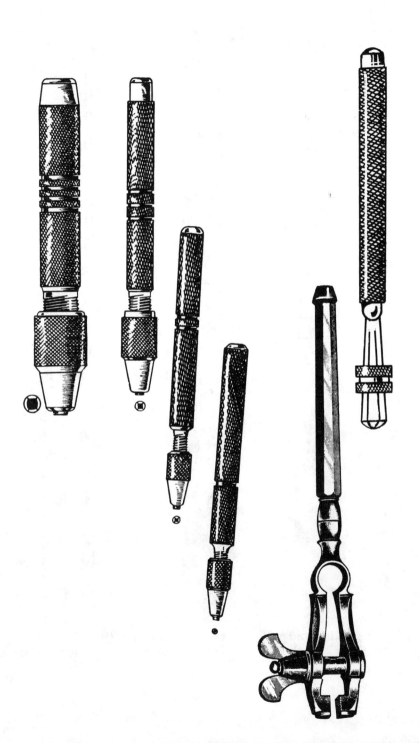

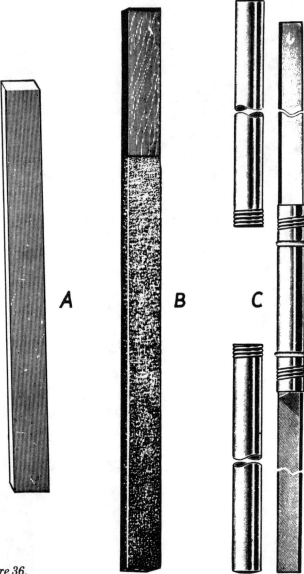

Figure 36.
Grinding and polishing tools.   **B—buff stick**
A—Arkansas slip   **C—pivot file and burnisher**

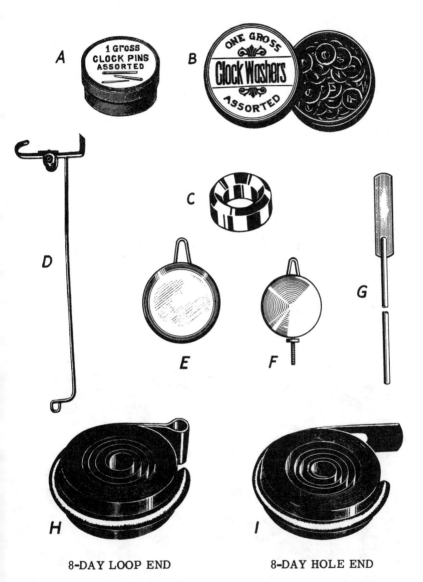

Figure 37.
Repair materials and clock parts

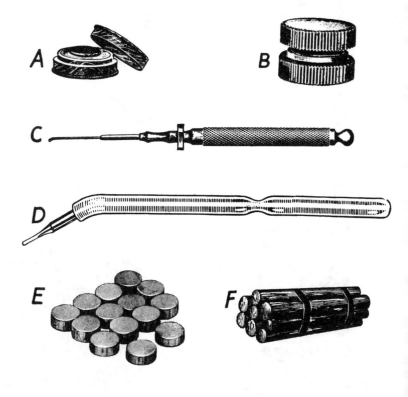

Figure 38.
Oil cups, oilers, and pithwood

In addition, they assumed that a complete dismantling of the movement was necessary. In today's clock servicing, however, the method has changed. Chemists have formulated special cleaning and rinsing solutions that are remarkably powerful. These were originally prepared for use in the ultrasonic watch-cleaning machine, but are equally useful for clocks. They clean the holes in the plates and the pivots of the wheels without the necessity of dismantling the movement.

The Workshop, the Bench, and Working Conditions

Having considered the tools and supplies needed for the clock hobbyist, the new clockmaker is ready to take a look at the shop itself. The bench (see Figure 39) should be fitted with a double-tube fluorescent lamp with a long adjustable arm so that the lamp can be moved into any position.

A bench plate, preferably of a green color, may be purchased from a tool and material dealer. This provides an excellent surface upon which to work. The apron of the bench should always be pulled out while the hobbyist is working. All repair benches have an apron and this equipment helps greatly in catching small parts that may be dropped or may flip out of the tweezers.

The modern repair bench is well supplied with drawer space. There should be a fixed place for every tool and every tool in its place. The only tools that should be left on the bench while working are eye loupes, tweezers, oil cups, oilers, small screwdrivers, and large pieces of button pithwood. Oilers may be inserted in the pithwood to keep the ends clean.

The lathe and motor are usually fitted on the left side of the bench top. Other tools used quite frequently should be arranged in drawers that are handy and accessible. Tools used less frequently may be placed in drawers not quite so handy but still within easy reach. In other words, the arrange-

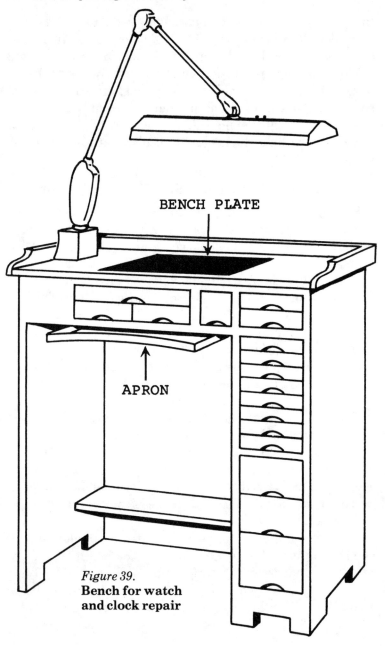

BENCH PLATE

APRON

Figure 39.
**Bench for watch
and clock repair**

ment of tools should be so planned as to make for pleasure **and** efficiency.

The hobbyist should be supplied with a posture chair, with cushion and an adjustable back. Since he will spend many hours at the bench, it is essential that satisfactory comfort be provided for peak efficiency. The type of chair is important if a tired and aching back is to be avoided.

The bench described above is suitable for the repair of shelf clocks and portable clocks. For the larger movements, however, such as are seen in grandfather clocks, a table where more space is available is to be preferred.

Now that the basic equipment has been enumerated and described, the hobbyist is ready to examine, clean, and repair a clock. This is the project with which the next chapter will deal.

7

The American
Striking Clock

There are so many different types and grades of clocks that an analysis of repair techniques must, of necessity, be concerned with essentials. The procedure outlined here, however, is entirely practical and sufficiently complete so as to cover the popular types of clocks commonly found in most homes and offices.

The analysis to follow deals with the cleaning, oiling, and repairing of the popular American shelf clock with counting-wheel striking mechanism. The 400-day clocks and the American alarm clocks are given separate consideration, inasmuch as these clocks are in a class by themselves. Repairs apart from cleaning and minor adjustments are given special consideration in Chapter 10.

Cleaning and Repairing the American Shelf Clock

Removing the movement from the case

Removing the movement from the case is generally a simple matter and, in most cases, too obvious even to call for an explanation. Most American shelf clocks in wooden cases are so designed that the hands and gong must be removed first. This done, remove the four screws that hold the movement in place and lift out the movement. A problem presents itself if one of the mainsprings is broken. In this case, wind the good spring fully and pull out the movement first on the side that contains the good spring. The movement will come out with some difficulty and it will take some caution to avoid bending the wheels.

Cleaning the movement

In time all clock movements become dirty. As the oil thickens it loses its lubricating quality and the clock eventually stops. This tendency to stop will occur more frequently during the cold winter months. This is a sure sign that the clock in question is in need of repair. Most people permit a clock to run too long without service. One should not wait until a clock stops. Instead, the pendulum clock should be cleaned after three or four years of running. This practice keeps the clock in good mechanical condition and avoids the wear that inevitably takes place when the oil thickens or dries up completely. Dirty oil acts like an abrasive and the holes in the plates and wheel pivots become badly worn.

If, after a thorough examination, the movement is found to be in good shape the movement may be cleaned without dismantling, except for a minor number of parts. For example, remove the escapement and crutch if the movement is similar to the model shown in Figure 4, p. 17. If the hobbyist finds it desirable, the pendulum rod, too, may be removed. However, if the movement is so designed that the escapement is fitted between the top and bottom plates, these parts are not removed.

Next, place the movement in a basin large enough to receive the movement. Pour in enough of the ultrasonic cleaning solution to cover the movement. Tip the basin in a manner that will splash the liquid in, around, and through the movement. If the movement is excessively dirty, use the brushes shown in Figure 33. In addition, allow the movement to remain in the solution for about fifteen minutes. Then remove the movement and place it over a heating lamp until dry. Now place the movement in another basin and pour in the ultrasonic rinsing solution as before and allow five to ten minutes for the rinsing procedure. Lastly, remove the movement from the rinse and dry over the heating lamp. The escapement, crutch, and pendulum rods are now cleaned, and replaced in the movement. The cleaning fluid itself should be saved because, unless the clock is extremely dirty, the solutions should satisfactorily clean ten or more clocks.

When the newly cleaned movement is completely dry, it is ready for the oiling procedure. Oil all the frictional areas with the clock oil. These include the point where the crutch surrounds the pendulum rod, and also every other tooth of the escape wheel. Fit the movement to the case and, for the present, leave off the hands. Wind the mainsprings fully; place the ball on the pendulum rod and start the clock. Observe its running and striking for a few days, and note particularly the beat. If out of beat, bend the crutch in such a manner (see "Adjusting the Crutch," p. 115) that the tick and the tock are equally spaced. Finally, replace the hands, check the time, regulate as needed.

Dismantling the movement

If the movement is in bad shape, obviously dismantling is required. Before taking the movement apart, however, examine the striking train and see if there are marks on the wheel and pinion which perform the function of locking and warning. If there are none, mark with a small screwdriver the points where the wheel gears with the pinion. If this is done, the striking mechanism may be assembled by noting the marks, and the striking action will remain correct—that is, if it was correct before dismantling.

The mainsprings are let down with double-end bench keys (see **D**, Figure 33, p. 75), but before doing so, place the spring clamps (see **H**, Figure 37, p. 81) over the springs. New mainsprings are enclosed in clamps, and several sizes of clamps may be acquired with the purchase of mainsprings. These should be saved for future use. Having placed the clamps around the springs, place the bench key of the correct size on the winding square and wind forward a fraction of a turn. This will make possible the release of the click as shown in Figure 1. While holding the bench key firmly between the thumb and fingers, lift the click away from the ratchet with a small screwdriver and allow the bench key to slip slowly between the thumb and fingers. It may be helpful to allow the click to return to the ratchet so as to get another firm grip on the bench key. Release the click a second time and let the mainspring down until it is completely unwound.

After the mainsprings are let down, examine the freedom of the pivots in the holes in the plates. If the pivot holes are badly worn, remove the top plate and lift out the timing train, keeping the several wheels together. Next remove the striking train and levers and keep these parts together and separate from the timing train. Slightly grooved pivots may be restored by using the pivot file and burnisher shown at **C** in Figure 36. Fit the wheel in the lathe and first apply the file as shown in Figure 40. Then finish the polishing with the burnisher.

Closing holes

Since the pressure on the pinion lies in the direction that the wheel is turning, it is seen that the wear takes place mostly on one side of the hole in the plate. The hole may be closed with a hollow center punch with half of the end

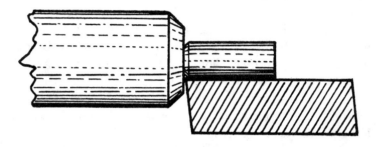

Figure 40.
Restoring pivot by use of pivot file

ground away. Rest the plate on an anvil and fit the punch to the side of the hole where the wear is greatest, then tap the punch sharply with a hammer. In this way the metal is worked toward the hole. After closing, the pivot should not go in the hole. Broach out the hole so that the pivot will fit freely. After closing all worn holes, assemble the wheels, one at a time, between the plates. Each wheel should spin freely. Then fit the complete train between the plates and note the freedom by turning the second wheel.

Oversize holes that are round may be closed with punches that are supplied in sets as shown in Figure 34. These are fitted with spring centers (pointed or round ends) that are placed in the holes. The procedure with spring center punches is similar to that described above.

Repairing the pallets

The pallets often need smoothing and polishing and this can be done with the Arkansas slip, a Number Three emery buff stick and the burnisher. The pallets are first ground with the Arkansas slip, with the movement of the tool lengthwise—that is, in the direction that the escape wheel teeth take in sliding along the pallet's impulse faces. Following this, polish with the Number Three emory buff stick and continue making the strokes lengthwise. Sliding the Arkansas slip and the buff stick lengthwise is important because a cross-wise movement would cut across the grain and the impulse face would than act as a file and so wear the teeth of the escape wheel. Now give the impulse faces of the pallets a final treatment with the burnisher. Again, the strokes should be made lengthwise.

After any remaining minor repairs have been performed, the entire mechanism should be cleaned in the manner explained earlier.

Assembling the movement

Rest the lower plate on a pasteboard box. First fit the mainsprings in place. Then fit the wheels, starting from the bottom of the movement and working toward the top. Next fit the striking levers. Follow with the top plate. Guide the center wheel (see Figure 41) to its hole in the top plate. Now press the top plate closer and, with the tweezers, guide the pivots of the larger wheels and lastly the smaller wheels to their respective holes in the top plate. Pin or screw down the top plate and test both the timing and striking trains for freedom. Wind the mainsprings, oil the movement and test the striking. Complete the job by fitting the movement into the case; place the ball on the pendulum rod, start the pendulum swinging, check the beat and then regulate as needed.

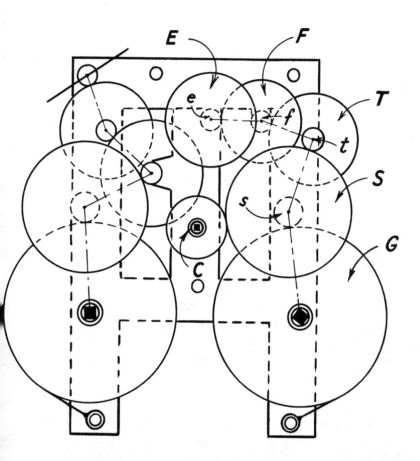

Figure 41.

Movement of American striking clock with top plate removed.

G—great or first wheel	**E—escape wheel**
S—second wheel	**s—second pinion**
C—center wheel	**t—third pinion**
T—third wheel	**f—fourth pinion**
F—fourth wheel	**e—escape pinion**

8

The 400-Day Clock

The 400-day clock is often referred to by other names, such as annual wind clock or anniversary clock. Instead of a pendulum it has the equivalent of a balance suspended at the end of a flattened wire. Designed as it is to run a year between windings, it will be noted (see Figure 42) that there are three intermediate wheels. The barrel makes one turn in 80 days and the balance makes 8 vibrations a minute, or 7½ seconds for each vibration. Exceptions may be found in some of the older models.

The escapement is of the dead-beat type. Some models are made with adjustable pallets as shown in Figure 43, while other manufacturers use pallets of one-piece construction.

Lying above the pallet frame is a round wire about 1½ inches long called the pallet pin, the upper end of which plays

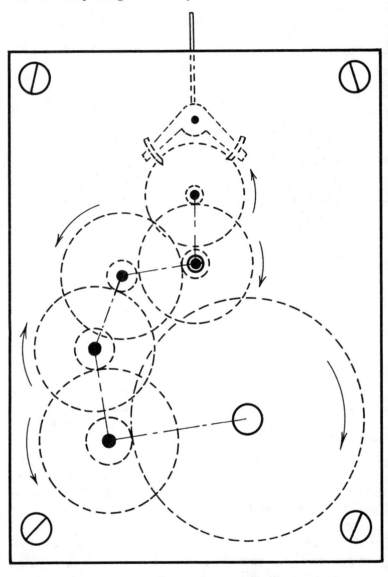

Figure 42.
The 400-day clock movement

freely in the slot of the fork **F** as shown in Figure 44. The fork is secured to the suspension spring **S**. When impulse is delivered to the fork, the suspension spring is twisted and the force is carried to the balance.

Repairing the 400-Day Clock

The analysis covering the cleaning of the American shelf clock applies also to the 400-day clock, but the adjusting of the escapement, the fitting of a new suspension spring, and the correcting of the beat present new problems. Readjusting of the pallets is rarely needed, but the correction of errors relative to the fork and pallet calls for comment.

Adjusting the beat

Suppose the clock is set up and running. This is the time to check the beat. Listen to the tick. The balance will continue its circular arc of motion after the tick, and the lock on the pallets will become deeper. This additional lock is called run and should be equal on both pallets. This may be noted by observing the distance covered by the balance after the tick. If it is equal for both clockwise and counterclockwise motions, the escapement is in beat.

If this additional motion is not equal, it may be made so by turning the block **B** that supports the upper end of the suspension spring. This piece is called the beat adjuster block and is fitted friction tight in the plate **P** (Figure 44). In some models, other designs are used. One, frequently seen, has a set screw that must be loosened to permit the turning of the beat adjuster block. However, before adjusting the block, it is well to inspect the suspension spring for bends or kinks that may lie above the fork. If the suspension spring appears in good condition, the beat error is corrected by turning the beat adjuster block in the direction that will bring the fork toward the side where the balance motion after the tick is shorter.

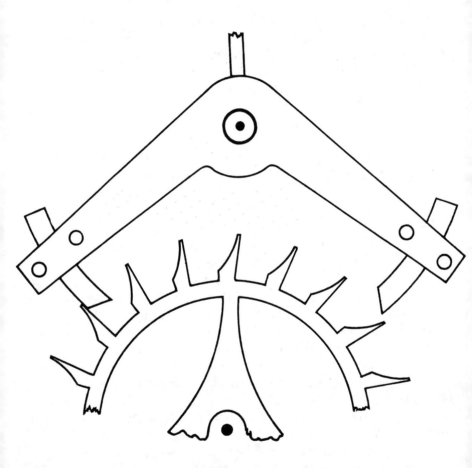

Figure 43.
Escapement used in 400-day clocks

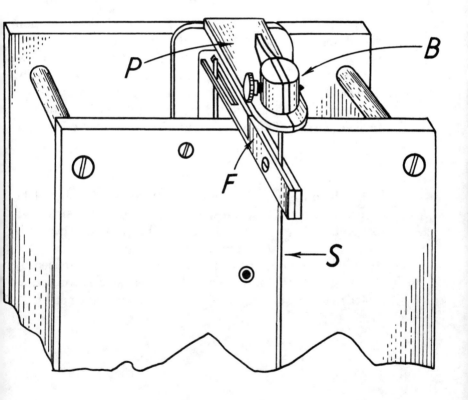

Figure 44.
Impulse assembly in 400-day clocks

Fitting a suspension spring

Suppose a clock for repair has a broken suspension spring. These may be obtained from tool and material dealers by sending the old spring as a sample. In fitting a new spring, first determine the required length. Then fit the end pieces and the fork. Secure the suspension assembly to the movement and set up the clock for running, check the beat and regulate to time.

Fluttering

Often, after fitting a new suspension spring, a malfunction called fluttering may be observed. This is a condition wherein the pallets receive and discharge several teeth of the escape wheel in the same time that the balance makes one turn. The clock may gain an hour or two in less than one hour. Fluttering can be eliminated by raising the fork on the suspension spring. This adjustment also tends to increase the balance arc of motion.

Balance motion

The balance arc of motion of 400-day clocks should be approximately 720 degrees, or 360 degrees for a single vibration. A shorter motion is permissible provided the run is adequate. A run that terminates too close to the tick could result in stopping. The run can sometimes be increased by bending the pallet pin toward the suspension spring or, as noted above, by raising the fork on the suspension spring.

9

The American
Alarm Clock

American alarm clocks, owing to their low cost, are not designed for extensive repairs. However, many of these clocks can be restored to good running condition at a minimum cost in time and money.

Repairing the American Alarm Clock

Dismantling the clock

The first act is to remove the time-winding and the alarm-winding keys that are located at the back of the case. Arrows on the case indicate the direction the keys are turned to wind. To remove, turn the keys in the opposite direction, as most keys are screwed on. In a few cases the keys are fitted friction tight to a square on the arbor. If this is the case, insert the jaws of cutting pliers under the keys and pry them off. Usually the movement may be removed from the case without disturbing the time-setting and the alarm-setting knobs, as the holes in the back plate are large enough to permit the knobs to go through.

With the movement out of the case, first remove the balance assembly. Remove the brass wedge to release the hairspring. With pliers, back out one of the threaded cups that carry the balance staff and remove the balance from the movement. Examine the pivots of the balance staff. The ends of the pivots should feel sharp when touched with the finger. If they are not sharp to the touch, they must be made so. Assuming that the correction is necessary, place the balance in a split chuck and grind the pivots to a sharp point in the lathe. For this purpose, use an Arkansas slip and finish with the steel burnisher.

The movement may now be cleaned without further dismantling, using, as already mentioned, the ultrasonic cleaning and rinsing solutions. The balance assembly should be cleaned separately, using two small jars for the cleaning and rinsing solutions. After the cleaning procedure, dry the movement and the balance over the heating lamp. After the movement and the balance are dry, complete the assembly procedure. Adjust the threaded steel cups to permit the needed end shake of the balance. Place the outer end of the hairspring between the regulator pins and secure the hairspring to the stud. Check the beat and make corrections if necessary.

Oiling the movement is next in order. Wind the time and alarm mainsprings fully and place clock oil on the coils. Oil the train wheel pivots, the balance pivots, and every other tooth of the escape wheel, and also the alarm pallets. The balance should now start with a brisk motion of not less than 450 degrees. A balance motion of 450 degrees means that the balance turns 225 degrees from its rest position, stops and returns to zero degrees, continues another 225 degrees and again stops and returns to zero degrees.

Adjusting the escapement

If the balance does not take a satisfactory motion, the escapement may be at fault. Turn the balance slowly until such time as to allow one tooth of the escape wheel to pass a pallet pin. The escape wheel will turn and another tooth will engage the opposite pallet pin. There should be a definite locking as shown in Figure 45. Failure to lock will show a pallet pin engaging the impulse face of a tooth. Using flat-nosed pliers, the correction consists in bending in, toward the escape wheel, the narrow projection of brass in which the pallet arbor is fitted. Satisfactory locking is thereby attained. This procedure puts the lever out of upright, since the opposite end usually has no such adjustment. This failure in alignment requires an additional adjustment; that of bending the lever to line up the fork with the impulse pin on the balance.

Adjusting the alarm

On the dial side of the movement lies a long flat spring with a bent-over end that reaches through a hole in the plate. This bent-over end locks or releases the alarm hammer. When turning the alarm setting knob, one can readily see what happens and the necessary adjustment can be made. If the bent-over end does not raise high enough to release the alarm hammer at the moment the alarm should ring, bend the long flat spring back slightly. If the alarm rings constantly, move it forward.

Replacing the hands

Turn the alarm setting knob slowly and stop at the instant the clock alarms. Fit the alarm hand to twelve o'clock.

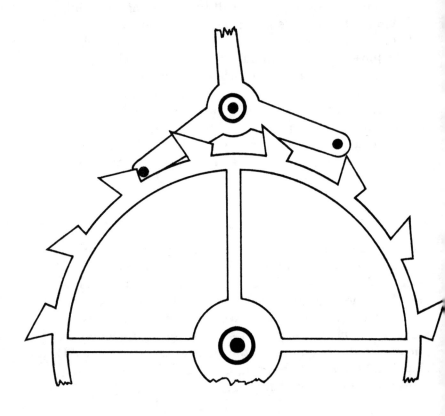

Figure 45.
Pin-pallet escapement

Likewise, fit the hour and minute hands to twelve o'clock. Next, turn the time setting knob clockwise and check the time when the clock alarms. If the alarm does not take place at twelve o'clock, readjust the hands until the clock alarms at the proper time.

10

Special
Repair Techniques

The analysis of clock servicing in Chapters 7, 8, and 9, covers the cleaning, minor repairs and adjustments and assumes that no particular difficulties are encountered. There are times, however, when it is not quite this simple, and this chapter will cover additional repair techniques that may be occasionally needed.

Mainspring Problems in Going-Barrel Clocks

The better quality clocks are fitted with barrels for the reception of the mainspring. In cleaning the clock the barrel cap should be removed. Pry off the cap by inserting a sharp instrument in the slot in the cap. If this method is found difficult, give the barrel arbor on the opposite end a sharp tap on some hard surface. The barrel cap will fly off. The mainspring need not be removed except in the case of breakage, or in a recently repaired clock that does not run the full time for which the clock was designed. In the latter case the problem could be a set mainspring (a horological term for loss of elasticity due to years of service). With either of the above problems a new mainspring is needed. The old mainspring can be removed and a new mainspring can be fitted by hand. This is rather difficult operation, however, and requires practice.

After fitting a new mainspring, note the space occupied by the spring. It should be similar to that shown in Figure 46. If found correct, place mainspring oil on the coils of the spring and replace the cap. If the empty space is larger than that shown in the illustration, the spring is too short and should be replaced with a longer spring. If there is less space than that shown in the illustration, the spring is too long and should be replaced with a shorter spring. In either case the clock would not run the full time for which it was designed.

Replacing a barrel hook

Often the barrel hook is pulled out and a new one needs to be made. To make a new hook, select a piece of steel wire and file the hook to the same shape as the old one. Then broach out the hole in the barrel to receive the new hook. Insert the hook in the hole and place the barrel on an anvil with the hook resting on the anvil. Using a round-end hammer, rivet the hook in place.

Repairing the end of the mainspring

The pear-shaped hole in the end of the mainspring occasionally pulls out. If the rest of the spring is in good condition, a new hole may be drilled in the end. Break off the outer end clear of the old hole and heat the end until it turns blue. Drill a new hole and file the opening to fit the hook in the barrel. Finish the end of the spring with a flat file, and bend the end beyond the hole to the same curve as the barrel wall.

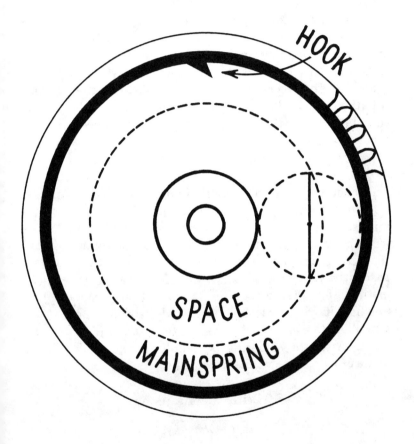

Figure 46.
Placement of mainspring in barrel

Repairing the Train

The worn pivot holes in good quality clocks are repaired by fitting bushings. Bushings, made up in assortments, may be purchased from tool and material dealers. In enlarging the hole in the plate to receive the bushing, it is often necessary to draw back the hole to its original position, since the wear is on one side. To do this, use a round needle file at the start. Then complete the job with a broach. Broach out the hole from the inside of the plate and make the hole just large enough to receive the bushing. Then drive the bushing into the plate and see to it that the bushing lies flush with the plate. Then, if needed, broach out the hole to fit the pivot.

Fitting teeth to a wheel

Broken teeth are occasionally found in clock movements. Fortunately the repair is not difficult. File a slot below the missing tooth to a depth equal to the height of the tooth. Select a piece of brass of the same thickness as the wheel and file it to fit the opening just prepared as shown in Figure 47. Insert the piece and solder with soft solder. A convenient method of soldering is to lay the wheel on a copper plate and hold it over an alcohol lamp until the solder melts. Caution should be exercised so as not to heat the pivots of the wheel. After the soldering is completed, the tooth is filed to the proper shape with a needle file.

Striking Mechanisms

Counting-wheel system

Failure of the counting-wheel system to strike properly is usually due to incorrect timing. Turn the minute hand to the hour and allow the striking to run slowly. When the counting lever drops into a deep slot in the counting wheel, the drop lever should lock in the cam wheel notch. If the cam wheel

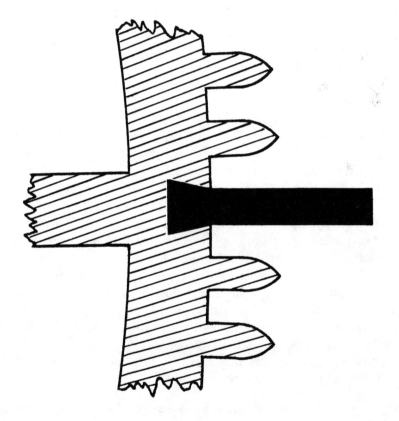

Figure 47.
Fitting new tooth to wheel

is not in the proper position, the drop lever cannot lock in the notch. Let the mainspring down and separate the plates on the striking side. Separate the cam wheel from the wheels next to it and turn the cam wheel so as to alter the timing. Assemble the striking train and try again.

Often the several levers need bending so that the drop lever will drop the necessary distance to engage the cam wheel.

Rack-and-snail system

Striking errors in the rack-and-snail system usually can be attributed to worn parts. The hole for the gathering pallet must fit closely. A large hole at this point causes the gathering pallet to hang up on the rack or fail to gather up the proper number of teeth.

Failure of the warning lever to drop at the hour could be due to a worn stop piece on the warning lever. Polishing off the wear mark will cause the striking to again function properly. Oiling the stop piece, too, would assist in causing the warning lever to drop at the hour.

The several levers of the rack-and-snail system should be free on the posts that carry them. This is mentioned because a faulty action may prove to be very puzzling, as a tight lever is not easily detected.

Another faulty adjustment, occasionally met with, is the lifting of the striking hammer when the clock warns. This results in a lot of useless run after the striking is completed. There should be very little run after the striking and no lifting of the striking hammer during the warning. Correct adjustment lessens the probability of the clock's failure to strike when the oil thickens.

Repairing Broken Hands

Hands often become broken near the center, but this presents no problem in the newer standard makes of clocks, as new hands can be purchased. However, hands for French clocks and antique clocks cannot be purchased and so the old ones must be repaired or new ones made completely. The following method for repairing old hands is suggested. In the French clocks a brass collet is fitted to the center of the hands. This collet must be removed from the broken piece. Now select a small piece of flat steel of about the same thickness as the hand. Drill a hole in it and enlarge the hole to fit the collet. File the steel piece to the shape shown in Figure 48. The width of the narrow part should be left a little wider than the hand and about half as long. Next fit the collet to the piece, making certain that the end points toward the 12 on the dial to insure correct striking. Then rivet securely. Now apply a little soldering flux and soft solder to the steel piece. With the soldering tweezers hold the steel piece and the broken hand together firmly and heat over the alcohol lamp until the solder melts. Finish by filing the surplus steel away. Clean the hand and, when fitted to the clock, the repaired hand will be found to look as good as the original.

Pendulums That Wobble

When a pendulum wobbles, something is wrong with the suspension assembly. Bends or kinks in the suspension spring could be the cause and, if these kinks can be flattened, the wobble should be eliminated. This is done with pivot straightening tweezers by applying pressure at the area of damage. If the wobble persists fit a new pendulum rod.

Another cause of wobbling may be found in those clocks which have a piece called a regulator bridge. This regulator unit is designed to be moved up or down on the suspension spring by applying a key to the square which extends through a hole in the dial. If the slot in the regulator bridge that receives the suspension spring is not parallel, the pendulum will wobble. Making the slot parallel should solve the problem.

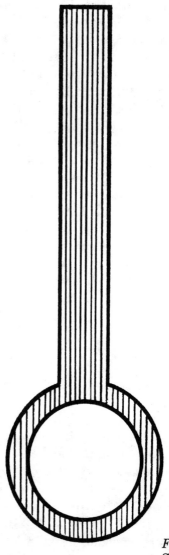

Figure 48.
**Step in repairing
old hand of clock**

Adjusting the Crutch

One more point should be observed when finally setting up the clock. See that the pendulum rod occupies the central area of the slot in the crutch. The adjustment can usually be done with the flat-nose pliers or the thumb and first finger. Suppose the slot area needs to be moved toward the movement. In this case bend the crutch downward near its top with the pliers. Again, suppose that the slot area needs to be moved away from the movement. In this case, grasp the slot with the thumb and first finger and pull it away from the movement. This adjustment may alter the beat, but a repeat of the beat correction should put the clock in satisfactory running condition.

Appendix

Tool and Material Dealers

BMS Materials, Box # 1, Pleasantville, N. Y. 10570.
Jules Borel and Company, 1110 Grand Avenue, Kansas City, Mo.
64106.
Stanley Donahue, Box # 53149, Houston, Texas 77052.
Friedman-Gessler Company, 315 West Fifth Street, Los Angeles,
Calif. 90013.
R. P. Gallien & Son, 220 West Fifth Street, Los Angeles, Calif.
90013.
A. J. Goldfarb, Inc., 251 West 30th Street, New York, N. Y. 10001.
Ralph Herman Clock House, 628 Coney Island Avenue, Brooklyn,
N. Y. 11218.
The Horolovar Company, Box # 44A, Bronxville, N. Y. 10708.
Kilb & Company, 623 North Second Street, Milwaukee, Wis.
53203.
S. LaRose, Inc., Greensboro, N. C. 27420.
Swartchild & Company, 22 Madison Street, Chicago, Ill. 60602.
United Tool and Material Company, 307 University Building, Den-
ver, Colo. 80202.

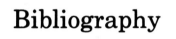

Bibliography

BOOKS

Bell, Harrington. *The Riddle of Time*. New York: The Viking Press, 1963.

Britten, F. J. *Old Clocks and Watches and Their Makers*. New York: Bonanza Books, 1956.

Bruton, Eric. *Dictionary of Clocks and Watches*. New York: Bonanza Books, 1963.

Cowen, Harrison J. *Time and its Measurement*. Cleveland: World Publishing Co., 1958.

deCarle, Donald. *Practical Clock Repairing*. London: N.G.A. Press, 1968.

Eckhardt, George H. *Pennsylvania Clocks and Clockmakers*. New York: Bonanza Books, 1950.

Fried, Henry B. *The Watchmakers' Manual*. Princeton, N. J.: D. Van Nostrand & Co., 1948.

Gazeley, W. J. *Watch & Clock Making and Repairing*. Princeton, N. J.: D. Van Nostrand & Co., 1953.

Goodrich, Ward. *The Modern Clock*. Chicago: North American, 1950.

Palmer, Brooks. *The Book of American Clocks.* New York: The Macmillan Co., 1950.

Rawlings, A. L. *The Science of Clocks and Watches.* New York: Pitman Publishing Co., 1948.

MAGAZINES

American Horologist and Jeweler, 2403 Campa Street, Denver, Colo. 80205.

Bulletin of the National Association of Watch and Clock Collectors, 335 North Third Street, Columbia, Penn. 17512.

Jewelers' Circular—Keystone, 56th and Chestnut Streets, Philadelphia, Penn. 19139.

Modern Jeweler, 1211 Walnut Street, Kansas City, Mo. 64106.

National Jeweler, 6 West 57 Street, New York City, N.Y. 10019.

The Northwestern Jeweler, 142 West Main Street, Albert Lea, Minn. 56007.

Pacific Goldsmith, 657 Mission Street, San Francisco, Calif. 94105.

Southern Jeweler, 75 Third Street, N. W., Atlanta, Ga. 30308.

Index

Index